The Apollo 13 Mission
Surviving an Explosion in Space

Helen Zelon

The Rosen Publishing Group's
PowerKids Press™
New York

For Gabriela Sasson, whose energy, vitality, and love for life outshine the Sun

Published in 2002 by The Rosen Publishing Group, Inc.
29 East 21st Street, New York, NY 10010

First Edition

Book Design: Michael de Guzman
Project Editors: Jennifer Landau, Jason Moring, Jennifer Quasha

Photo Credits: pp. 4, 7, 16 © Bettmann/CORBIS; pp. 8, 15, 19 © PHOTRI-MICROSTOCK; p. 11 Digital image © 1996 CORBIS; p. 12 © UPI/CORBIS-BETTMANN; p. 20 (diagram) © UPI/CORBIS-BETTMANN and (capsule) Bettmann-CORBIS.

Zelon, Helen.
The Apollo 13 mission : surviving an explosion in space / Helen Zelon.
    p. cm. — (Space missions)
Includes bibliographical references and index.
ISBN 0-8239-5773-X (library binding)
1. Apollo 13 (Spacecraft)—Juvenile literature. 2. Project Apollo (U.S.)—Juvenile literature. 3. Space vehicle accidents—Juvenile literature.
[1. Apollo 13 (Spacecraft) 2. Project Apollo (U.S.) 3. Space vehicle accidents] I. Title. II. Series.
TL789.8.U6 A69424 2002
629.45'4—dc21
                                                                                              00-013042

Manufactured in the United States of America

Contents

Exploring the Moon

Apollo 13 was the U.S. space program's third mission to the Moon. The Apollo 13 astronauts were James Lovell, John Swigert, and Fred Haise. Their spacecraft had three parts, the **command module**, the **service module**, and the **lunar module**. The lunar module is the only part of the spacecraft that actually lands on the Moon.

Most of the supplies for Apollo 13 were stored in the service module. This area held oxygen for the astronauts to breathe in space. It also held special **fuel cells** that were used instead of heavy batteries. The cells blended liquid hydrogen and oxygen to make electricity. This mixture gave off an important **by-product**, fresh water. The water was for the astronauts to drink in space. It also was used to cool the spacecraft's equipment during its long flight.

← *This is a photograph of Apollo 13 astronauts (from left to right) Fred Haise, James Lovell, and John Swigert.*

False Alarm

On April 11, 1970, at 2:13 P.M. on a bright spring afternoon, Apollo 13 lifted off into the sky. Five minutes after the **launch**, the astronauts felt the spacecraft shake. Worried scientists and engineers at **Mission Control** saw that one of the five rocket engines had shut down 2 minutes early. Mission Control made sure that the other four engines burned a few minutes longer. This extra time helped Apollo 13 get into **orbit** around Earth.

The Apollo 13 crew was relieved to be safe in Earth's orbit. The engineers and scientists on the ground were proud that they had solved the problem. Astronauts Lovell, Haise, and Swigert were ready to go to the Moon. No one knew that an even bigger problem lay ahead.

This photograph shows the Apollo 13 spacecraft as it headed off into space. ➤

HOUR   MINUTE   SECOND

Smooth Sailing

For the next two days, the Apollo 13 astronauts spent most of their time in the command module, where they conducted experiments and prepared for their Moon landing. On the evening news on April 13, the astronauts sent down images of their spacecraft to millions of television viewers on Earth. There is no **gravity** in space, so the astronauts didn't walk from area to area. Instead, they floated! The astronauts floated through a small tunnel to pass from the command module to the lunar module. The lunar module was much smaller and much less comfortable than the command module. The lunar module was meant to supply food and shelter for two people over two days. That was the amount of time the Apollo 13 astronauts would be on the Moon's surface.

← *These are images of astronauts Fred Haise* (top) *and James Lovell* (bottom), *sent down to Earth from space.*

Explosion in Space

All was quiet on Apollo 13. The astronauts were getting ready to go to sleep. Suddenly, at 9:08 P.M., they heard a loud BOOM! They felt a huge jolt. They saw a stream of gas rushing out into the sky. Mission Control specialists realized that oxygen was escaping from the spacecraft. Without oxygen, the astronauts would not be able to breathe! They discovered that a large explosion had taken place in the service module. The service module held Apollo's two oxygen tanks and the fuel cells that made electrical power. A spark in one oxygen tank had caused the explosion. The fire had ruined most of the command module's electrical **systems**. The command module had no light, water, or power. The astronauts were trapped 200,000 miles (321,869 km) above Earth.

This is a picture of one of the astronauts asleep aboard the Apollo 13 spacecraft. The explosion took place just as the astronauts were getting ready for bed.

A Lifeboat in Orbit

The astronauts decided to leave the command module for the lunar module, which had not been destroyed in the explosion. The command module had no oxygen and just a small bit of fuel. This fuel would be used to return the astronauts to Earth because the lunar module could not make it through **re-entry** into Earth's atmosphere. The command module had a **heat shield**, which would protect the astronauts during re-entry.

Near midnight the astronauts left the command module for the lunar module. Its two-day supply of oxygen and water could keep two astronauts alive during a lunar exploration. It would take four days to get the three Apollo 13 astronauts back to Earth from space. The supplies in the lunar module would have to last for those four days.

← *This is a drawing of the command module and lunar module of Apollo 13.*

The Long Way Home

The Apollo 13 astronauts sat close together inside the lunar module. The command module was dark and empty, but it had to be protected. It would be needed to return the astronauts to Earth. Apollo 13 was too far from Earth to turn around and come home right away. The astronauts had to continue their path around the Moon. They didn't eat or sleep the night of April 13 as they worked on their broken spacecraft. They used the fuel cells in the lunar module for power. Extra electricity was saved in batteries that could be used for the command module's return to Earth. The command module's **navigation** system had been ruined in the explosion. The astronauts used the Sun to make sure they were headed the right way. The spacecraft made its way around the Moon and back toward Earth.

This is a drawing of the inside of the lunar module, where the Apollo 13 astronauts spent much of their time in space. ➡

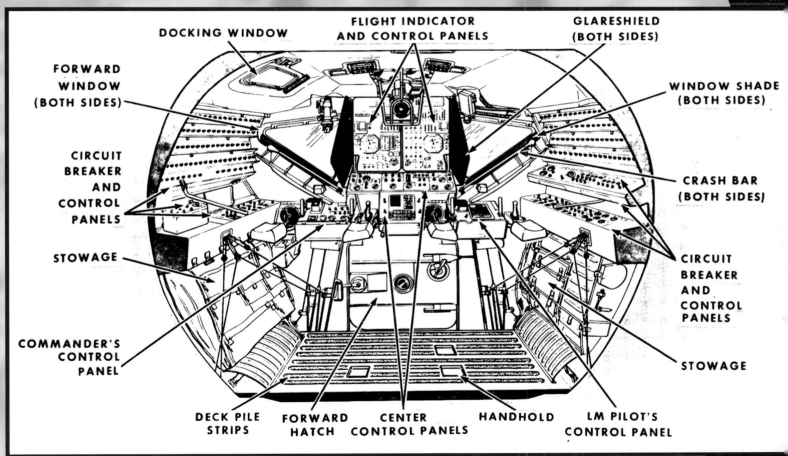

DOCKING WINDOW

FLIGHT INDICATOR AND CONTROL PANELS

GLARESHIELD (BOTH SIDES)

FORWARD WINDOW (BOTH SIDES)

WINDOW SHADE (BOTH SIDES)

CIRCUIT BREAKER AND CONTROL PANELS

CRASH BAR (BOTH SIDES)

STOWAGE

CIRCUIT BREAKER AND CONTROL PANELS

COMMANDER'S CONTROL PANEL

STOWAGE

DECK PILE STRIPS

FORWARD HATCH

CENTER CONTROL PANELS

HANDHOLD

LM PILOT'S CONTROL PANEL

Miserable and Cold

Astronauts Swigert, Lovell, and Haise now lived in the lunar module. The command module sat waiting for their return to Earth's atmosphere four days later. The astronauts had to save energy to make the lunar module's fuel last long enough for their return home. To save fuel, they did not use heat. The temperature in the lunar module dropped to 38 degrees Fahrenheit (3.3 degrees C), only a little warmer than freezing. The astronauts were hungry and thirsty. To save water, they could drink only 6 ounces (18 cl) of water a day. That's less than one juice box! Most of the water was needed to cool the equipment on the spacecraft. Thirty-six hours after the explosion, Mission Control experts ordered the astronauts to get some sleep. While they rested, Mission Control worked to bring home the Apollo 13 crew safely.

The scientists and engineers of Mission Control, shown in this photograph, worked hard to save the astronauts of Apollo 13.

Help from the Ground

The Apollo 13 astronauts felt dizzy and sick. They realized that the lunar module was filling up with **carbon dioxide**. When we breathe, we breathe in oxygen and breathe out carbon dioxide. Carbon dioxide is poisonous. Without fresh air to breathe, the astronauts would die. In space, carbon dioxide is stored in **canisters** that seal up the gas. The astronauts had filled up the lunar module's canisters. Extra canisters in the command module did not fit into the lunar module. Using plastic bags, cardboard notebook covers, and electrical tape, Mission Control figured out a way to make the command module's canisters work in the lunar module. Thanks to help from Mission Control, the astronauts stayed alive.

In this photograph, John Swigert holds the machine used to empty the lunar module of carbon dioxide. ➔

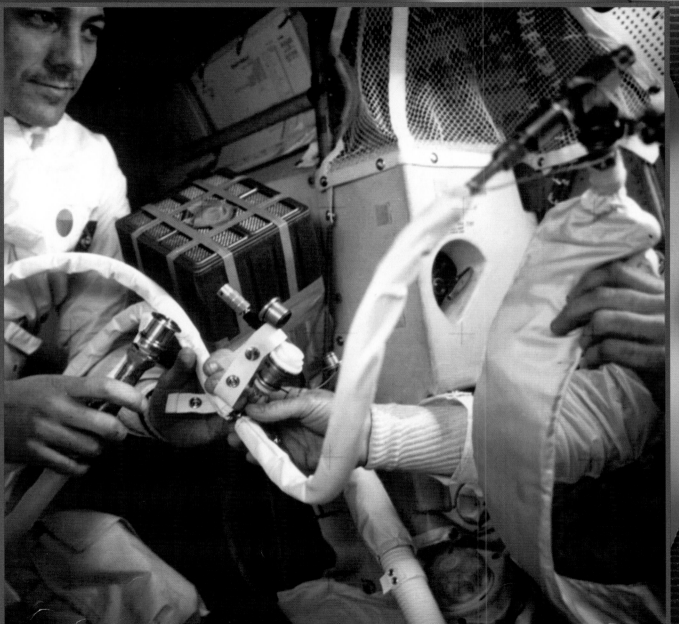

APOLLO 13 VOYAGE

APOLLO 3RD STAGE
MOONBOUND

① 4/11

4/13 ②

4/14 ③

LM ENGINE
FIRES 2ND
TIME

4/14 ④

OXYGEN TANK
RUPTURES IN
SERVICE MODULE

LM ENGINE
FIRES

⑥ LAUNCH
CAPE KENNEDY

⑤ 4/17

LUNAR MODULE JETTISONS SERVICE MODULE

⑤ SM   LM

REENTRY ⑥

Power Up

After four days and nights, the astronauts got ready to leave the lunar module. They opened the lunar module **hatch** and floated into the dark command module. Moisture had collected on the walls and instrument panels in the days since the explosion. The astronauts worried that the wires behind the panels were wet, too. A single drop of water could destroy important electrical systems. A spark could cause another fire. System by system, the astronauts powered up the command module. Everything worked. They had enough electrical power to operate the spacecraft and enough fuel to return to Earth. They did not know how much harm had been done to other parts of their spacecraft, however. They weren't safe on the ground yet, but they were coming home.

Top: *The drawing shows the flight of Apollo 13, from the launch to the* ← *explosion and the astronauts' return to Earth.* Right: *This is a photograph of the command module after it had splashed down in the Pacific Ocean.*

A Safe Return

The astronauts buckled themselves into the command module's seats. Before they returned to Earth's atmosphere, they released the lunar module into space. It had never landed on the Moon, but it had been their home in space. The astronauts also released the service module before re-entry. It floated away from the command module and down toward Earth.

The astronauts plunged back into Earth's atmosphere. On April 17, at 1:07 P.M., after 5 days, 22 hours, and 54 minutes in space, the Apollo 13 command module splashed down in the Pacific Ocean, near the Polynesian islands of Fiji and Tonga. Sailors on the U.S. ship *Iwo Jima* were waiting to rescue the astronauts. The tired astronauts were thrilled to be safe on Earth at last.

Glossary

by-product (BY-prah-dukt) A result of an industrial or a biological process.

canisters (KA-nih-sturz) Containers for holding objects or substances.

carbon dioxide (KAR-bun dy-OK-syd) A heavy, colorless gas exhaled by animals.

command module (kuh-MAND MAH-jool) The area of a space capsule where astronauts live and work.

fuel cells (FYOOL SELZ) An object that changes the chemical energy of things into electrical energy.

gravity (GRA-vuh-tee) The force that pulls things toward the center of Earth.

hatch (HACH) The doorlike opening in a spacecraft.

heat shield (HEET SHEELD) An area on a flat base of a command module that protects a space capsule during re-entry.

launch (LAWNCH) To push out or put a spacecraft into the air.

lunar module (LOO-ner MAH-jool) The part of a spacecraft that is meant to land on the Moon.

Mission Control (MIH-shun kun-TROL) A group of scientists and engineers who guide space travel from the ground.

navigation (nah-vih-GAY-shun) The science of getting from place to place, using maps, charts, and other objects.

orbit (OR-bit) The path one body makes around another, usually larger, body.

re-entry (ree-EN-tree) The return to Earth's atmosphere from space.

service module (SUR-vis MAH-jool) The part of the spacecraft underneath the command module and above the lunar module.

systems (SIS-tums) Groups of objects that are used for a special purpose.

Index

Web Sites

To find out more about the Apollo 13 mission and spaceflight, check out these
   Web sites:
     http://spaceflight.nasa.gov
       www.jsc.nasa.gov
         www.pbs.org/wgbh/nova/tothemoon